# FUN FACTS ABOUT SCIENCE:
# 240+ TRIVIA
# QUESTIONS
## FOR KIDS, AGES 8-12

### BY ELIZABETH JAMES

ISBN 979-8-9891101-5-5
(paperback)

Published in the United States by Big Heart Books.

BIG HEART
books

# TABLE OF CONTENTS

# EXCLUSIVE FREEBIE!

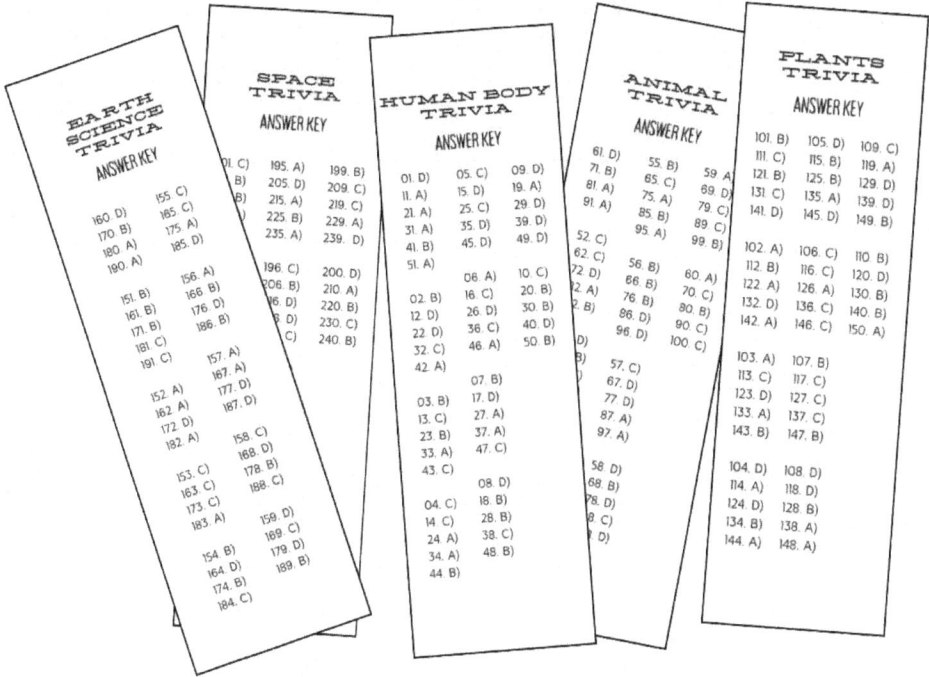

## EARTH SCIENCE TRIVIA
### ANSWER KEY

160. D)
170. B)
180. A)
190. A)

155. C)
185. C)
175. A)
185. D)

151. B)
161. B)
171. B)
181. C)
191. C)

156. A)
166. B)
176. D)
186. B)

152. A)
162. A)
172. D)
182. D)

157. A)
167. A)
177. D)
187. D)

153. C)
163. C)
173. C)
183. C)

158. C)
168. D)
178. B)
188. C)

154. B)
164. D)
174. B)
184. C)

159. D)
169. C)
179. D)
189. C)

## SPACE TRIVIA
### ANSWER KEY

01. C)
B)
B)

195. A)
205. D)
215. A)
225. B)
235. A)

199. B)
209. C)
219. C)
229. A)
239. D)

196. C)
206. B)
6. D)
8. D)
C)

200. D)
210. A)
220. B)
230. C)
240. B)

## HUMAN BODY TRIVIA
### ANSWER KEY

01. D)
11. A)
21. A)
31. A)
41. B)
51. A)

05. C)
15. D)
25. C)
35. D)
45. D)

09. D)
19. A)
29. D)
39. D)
49. D)

02. B)
12. D)
22. D)
32. C)
42. A)

06. A)
16. C)
26. D)
36. C)
46. A)

10. C)
20. B)
30. B)
40. D)
50. B)

03. B)
13. C)
23. B)
33. A)
43. C)

07. B)
17. D)
27. A)
37. A)
47. C)

04. C)
14. C)
24. A)
34. A)
44. B)

08. D)
18. B)
28. B)
38. C)
48. B)

## ANIMAL TRIVIA
### ANSWER KEY

61. D)
71. B)
81. A)
91. A)

55. B)
65. C)
75. A)
85. B)
95. A)

59. A)
89. D)
79. C)
89. C)
99. B)

52. C)
72. D)
2. A)
2. B)

56. B)
66. B)
76. B)
86. B)
96. D)

80. A)
70. C)
80. B)
90. C)
100. C)

D)
B)

57. C)
67. D)
77. D)
87. A)
97. A)

58. D)
68. B)
78. D)
8. C)
D)

## PLANTS TRIVIA
### ANSWER KEY

101. B)
111. C)
121. B)
131. C)
141. D)

105. D)
115. B)
125. B)
135. A)
145. D)

109. C)
119. A)
129. D)
139. D)
149. B)

102. A)
112. B)
122. A)
132. D)
142. A)

106. D)
116. C)
126. A)
136. C)
146. C)

110. B)
120. D)
130. B)
140. B)
150. A)

103. A)
113. C)
123. C)
133. A)
143. B)

107. B)
117. C)
127. C)
137. C)
147. B)

104. D)
114. A)
124. D)
134. B)
144. A)

108. D)
118. D)
128. B)
138. A)
148. A)

## GET ALL THE ANSWERS AT YOUR FINGERTIPS!

## DOWNLOAD FREE PRINTABLE ANSWER KEY BOOKMARKS

### WHEN YOU SIGN UP FOR EMAILS!

Visit elizabethjameswrites.com/scienceanswers

# CHECK OUT MORE TRIVIA FUN!

## 240+ TRIVIA QUESTIONS ABOUT ALL YOUR FAVORITE HOLIDAYS!

# INTRODUCTION

Science is all around us—from the plants and animals in our backyard to the stars up above at night. The more you learn about it, the more fascinating it all becomes!

That's what this book is about: celebrating the science we find in the world around us. And what better way to do that than by making learning about science fun?!

That's why we've compiled the most interesting facts about five of our favorite science subjects and turned them into interactive trivia questions you can use to unlock your knowledge about science!

The questions are organized into five different science topics—the human body, animals, plants, earth science and space—and include multiple-choice answers that make it easy for everyone to play and participate.

Plus, we've also included 8 different game challenges to make the learning even more fun. They are ultra-portable and involve little set-up, so you can take them with you on-the-go, whether you're playing solo or with others as you challenge your scientific knowledge.

So, pack your bags, and let your science trivia adventures begin!

# HUMAN BODY TRIVIA

1. In the United States, the average person lives to be about 78 years old. Jeanne Louise Calment holds the world record for living the longest. She was born on February 21, 1875 in France and lived through the building of the Eiffel Tower and met Vincent van Gogh. How old was she when she died?
A) 107 years old
B) 112 years old
C) 117 years old
D) 122 years old


2. Humans need water to live. How long can the average person go without drinking any water?
A) 3 hours
B) 3 days
C) 3 weeks
D) 3 months

ANSWERS ON PAGE 103

**3. Sleep is vital for humans. Three of the following are reasons why this is true; which of the following statements is *not* true about why sleep is important?**
A) Sleep allows your heart to rest.
B) Sleep allows your brain to go dormant.
C) Sleep helps your body regulate your blood sugar.
D) Sleep allows your body to perform necessary repairs.

**4. Which of the following is the largest organ in the human body?**
A) Brain
B) Lungs
C) Skin
D) Large intestines

**5. The largest muscles in the human body are called your "gluteus maximus." Where in the body are these muscles found?**
A) Leg
B) Chest
C) Butt
D) Stomach

**6. Which of the following is the body part that attaches your muscles to your bones?**
A) Tendons
B) Neurons
C) Nephrons
D) Stirrups

## 7. Which of the following is the longest organ in the human body?

A) Colon
B) Small intestine
C) Large intestine
D) Esophagus

## 8. What are the tubes called that carry blood to the heart?

A) Arteries
B) Nerves
C) Tendons
D) Veins

## 9. Most of the human body stops growing by the time you reach adulthood. However, there are two body that never stop growing. Which of the following are those two body parts?

A) Heart and lungs
B) Teeth and bones
C) Stomach and intestines
D) Nose and ears

## 10. Where is the smallest bone in the human body found?

A) Little toe
B) Baby tooth
C) Inside your ear
D) In your spine

ANSWERS ON PAGE 103

**11. There are 33 bones in the human spine. What are these bones called?**
A) Vertebrae
B) Ribs
C) Cartilage
D) Joints

**12. The ear is divided into the "outer ear," "middle ear," and the "inner ear." What separates the outer and middle parts?**
A) Cochlea
B) Cartilage
C) Temporal bone
D) Eardrum

**13. What is the average temperature of the human body in Fahrenheit degrees?**
A) 86.2 degrees
B) 93.3 degrees
C) 98.6 degrees
D) 100 degrees

**14. The average adult has 32 teeth. How many teeth did the person who set the record for the most teeth have?**
A) 35
B) 36
C) 37
D) 38

**15. Most babies get their first teeth when they are around 4 to 7 months old. However, in 1990, Sean Keaney was born in the U.K. with the record for the most teeth at birth. How many teeth did he have when he was born?**

A) 6

B) 8

C) 10

D) 12

**16. In the United States, the average height for a man is almost 6 feet tall. The world's tallest man ever was American Robert Wadlow when he was measured in 1940. How tall was he?**

A) Almost 7 feet tall

B) Almost 8 feet tall

C) Almost 9 feet tall

D) Almost 10 feet tall

**17. An adult human has 206 bones in its skeleton. However, when babies are born, they actually have *more* bones than that. Over time, these extra bones grow together and combine. About how many bones does a baby usually have when it's born?**

A) 225

B) 250

C) 275

D) 300

ANSWERS ON PAGE 103

**18. Which of the following is the fattiest organ in the human body?**
A) Kidney
B) Brain
C) Stomach
D) Heart

**19. When a woman is pregnant, she grows an extra organ to sustain the baby as it develops. Which organ is it?**
A) Placenta
B) Kidney
C) Uterus
D) Appendix

**20. One organ in the body specializes in filtering your body's blood, including removing old and damaged red blood cells and detecting bacteria or viruses in your blood. What is this organ called?**
A) Heart
B) Spleen
C) Appendix
D) Kidneys

**21. What job in the body does your diaphragm help with?**
A) Breathing
B) Digesting food
C) Dreaming
D) Sweating

**22. What are the bones called that can be found in your chest, protecting your lungs and heart?**
A) Skull
B) Spine
C) Phalanges
D) Ribs

**23. How many lungs do humans have?**
A) 1
B) 2
C) 3
D) 4

**24. Humans breathe in air to get the gas oxygen, which our bodies need. What is the name of the gas that we breathe out, which our bodies do *not* need?**
A) Carbon dioxide
B) Hydrogen
C) Nitrogen
D) Carbon monoxide

**25. A cheetah can run 62 miles per hour. How fast can a sneeze travel?**
A) 24 miles per hour
B) 45 miles per hour
C) 68 miles per hour
D) 88 miles per hour

ANSWERS ON PAGE 103

26. Your iris is the colored part of your eyeball, which can be different colors. Which of the following is *not* one of the iris colors people might be born with?
A) Amber
B) Gray
C) Green
D) Yellow

27. Which eye color is the most common eye color around the world?
A) Brown
B) Blue
C) Green
D) Hazel

28. Which organ pumps blood through the human body?
A) Lungs
B) Heart
C) Stomach
D) Brain

29. Blood is like a highway in the human body, moving many things around. Three of the following are things blood transports throughout the body; which of the following does blood *not* transport?
A) Oxygen
B) Nutrients from food
C) Hormones
D) Water

## 30. What color is human blood inside our veins?
A) Blue
B) Red
C) Clear
D) Purple

## 31. The human tongue is divided into four separate regions that detect different kinds of tastes. Which of the following is *not* one of the taste regions?
A) Sticky
B) Sour
C) Salty
D) Sweet

## 32. Ears do more than just help you hear. Which of the following is also one of their functions?
A) Provide a sense of direction
B) Dream at night
C) Keep your balance
D) Fight off colds

## 33. The throat contains two tubes; one carries food and drink when we eat, and one carries air to help us breathe. What is the tube for breathing called?
A) Trachea
B) Esophagus
C) Sphincter
D) Pancreas

ANSWERS ON PAGE 103

**34. After you eat, your body must get rid of any extra waste it doesn't use. This is what comes out when you poop. What is the proper name for this kind of waste?**
A) Feces
B) Ureters
C) Kidney stones
D) Fibers

**35. Poop is made out of bacteria and undigested food. But it consists mostly of which of the following:**
A) Blood
B) Skin
C) Tissues
D) Water

**36. What is the main function of kidneys in the human body?**
A) To clean your blood
B) To digest your food
C) To produce urine
D) To help fight off infections

**37. What is the substance called in the human body that gives color to our skin, hair and nails?**
A) Melanin
B) Keratin
C) Collagen
D) Blood vessels

38. There is one nerve that runs through the body that scientists have discovered helps regulate many of the body's functions—including breathing, digesting food, heart beat and more—that also can help you feel calm. What is this nerve called?
A) Optic nerve
B) Ulner nerve
C) Vagus nerve
D) Radial nerve

39. There are seven major systems found in the human body that work together to make it function. Which of the following is the system associated with moving blood throughout the body?
A) Skeletal system
B) Nervous system
C) Respiratory system
D) Circulatory system

40. The respiratory system is how your body breathes in air, and multiple body parts make up this system. Three of the following body parts are part of this respiratory system. Which of the following body parts is *not* a part of the respiratory system?
A) Mouth
B) Lungs
C) Diaphragm
D) Stomach

ANSWERS ON PAGE 103

41. Which of the major systems in the human body is associated with defending your body against germs and infections to help you stay healthy?
A) Digestive system
B) Immune system
C) Renal system
D) Endocrine system

42. One of the key systems in the human body is the "immune system." What does "immune" mean?
A) Protected
B) Diseases
C) Hungry
D) Tired

43. The brain, spinal cord and nerves work together to form which major system in the human body that is your body's way of sending messages throughout itself?
A) Endocrine system
B) Lymphatic system
C) Nervous system
D) Cardiovascular system

44. The human skull is made up of 21 bones. What is the name of the main bone that surrounds and protects the brain?
A) Ulna
B) Cranium
C) Femur
D) Scapula

**45. Lymph nodes are part of the immune system and help filter out germs in the body. They can be found in various parts of the human body. Three of the following are places in the human body where lymph nodes can be found. Which of the following is *not* one of the places where you will find lymph nodes?**
A) Neck
B) Groin
C) Armpit
D) Forehead

**46. The trachea is a part of the respiratory system that helps you breathe. What is another name for this tube, found in your throat?**
A) Windpipe
B) Esophagus
C) Intestines
D) Appendix

**47. Polydactylism can be a genetic condition that runs in families, resulting in people being born with extra fingers and toes. This condition resulted in a person setting the record for being born with the most fingers and toes: While most people are born with 10 fingers and 10 toes, how many of each did the person who set the record have?**
A) 12 fingers, 12 toes
B) 13 fingers, 18 toes
C) 14 fingers, 20 toes
D) 15 fingers, 14 toes

ANSWERS ON PAGE 103

**48. Three of the following are true about sweat glands in the human body. Which of the following is *not* true about our sweat glands?**
A)  They help regulate our body temperature.
B)  They help sense pain.
C)  They help eliminate water, salt and other wastes from the blood.
D)  They are found all over the body.


**49. Farting (also called "flatulence") is what happens when our body needs to get rid of bacteria in the large intestine that can't be metabolized. What would happen if we didn't fart?**
A)  We would burp.
B)  We would sneeze.
C)  We would have diarrhea.
D)  We would explode.


**50. The Guinness World Record is four feet for which of the following longest hair records?**
A)  The longest bangs
B)  The longest mohawk spike
C)  The longest beard
D)  The longest armpit hair

**51. Hirsutism is the condition when women grow excessive amounts of body hair, including on their face, which can result in growing a beard. This is usually caused by what?**

A) Their bodies probably produce excess male hormones.

B) They have probably eaten too much beef jerky.

C) Their bodies probably grew it to keep them warm.

D) They probably inherited more genes from their dad than their mom.

ANSWERS ON PAGE 103

# ANIMAL TRIVIA

52. Mammals are warm-blooded animals that feed their babies with milk and have some sort of hair or fur on their bodies. Three of following animals are mammals; which of the following is *not*?
A) Kangaroo
B) Sperm whale
C) Bald Eagle
D) Human

53. Most mammals give birth to their babies. Which of the following mammals actually lays eggs (known as a "monotreme")?
A) Dolphin
B) Orca
C) Bat
D) Platypus

ANSWERS ON PAGE 103

**54. What is the largest land mammal in the world?**
A) Elephant
B) Rhinoceros
C) Giraffe
D) Polar bear

**55. Most mammals walk on land or swim. While some mammals can jump and glide in the air, only one mammal is able to actually fly. Which one is it?**
A) Flying squirrel
B) Bat
C) Sugar gliders
D) Flying lemur

**56. Not all birds can fly. Three of the following birds cannot fly. Which is the only one in the following list that _can_ fly?**
A) Ostrich
B) Chicken
C) Penguin
D) Emu

**57. What is the largest species of lizard?**
A) Blue iguana
B) Crocodile monitor
C) Komodo dragon
D) Gila monster

**58. Groups of animals are sometimes given a unique name, such as a "school" of fish. Three of the group names below are true; which of the following is *not* a real name for a specific group of animals?**
A) A "charm" of goldfinches
B) A "dazzle" of zebras
C) A "crash" of rhinos
D) A "spot" of ladybugs

**59. Animals that are primarily active at night are known as "nocturnal" animals. Three of the following are all nocturnal animals. Which of the following is *not* a nocturnal animal?**
A) Squirrel
B) Opossum
C) Hedgehog
D) Hamster

**60. The kookaburra is a bird from Australia. What is it known for?**
A) Making a strange call that sounds like laughter
B) Bright feathers that look like a mohawk
C) Spitting at people passing by
D) Stealing cookies at picnics

**61. What is the largest species of shark?**
A) Cow shark
B) Bull shark
C) Great white shark
D) Whale shark

ANSWERS ON PAGE 103

**62. Three of the following animals have transparent skin. Which of the following does *not*?**
A) X-ray fish
B) Glass frog
C) Chameleon
D) Jellyfish

**63. A cheetah is the fastest land mammal, capable of running at 64 miles per hour. (That's about as fast as a car on the highway!) However, it can only run that fast for 20 seconds, and then it must slow down. Which of the following mammals can run the fastest for the longest time—clocking in at 53 miles per hour for miles and miles?**
A) Greyhound dog
B) Pronghorn antelope
C) Giraffe
D) Race horse

**64. What is the fastest fish in the world?**
A) Sailfish
B) Swordfish
C) Barracuda
D) Four-winged flying fish

**65. Which animal below is able to reach speeds of 200 miles per hour?**
A) Swift moth
B) Bullet ant
C) Peregrine falcon
D) Hummingbird

66. Some animals can change their color to camouflage with their environment, even if it's just for a specific season. Three of the following can change color to some degree. Which of the following does *not* change its colors?
A) Alaskan hare
B) Parrot
C) Peron's tree frog
D) Chameleon

67. Which of the following mammals can hold its breath for up to 90 minutes?
A) Polar bear
B) Orca
C) Dolphin
D) Sperm whale

68. Which of the following is the world's smallest species of bird?
A) Cinnamon hummingbird
B) Bee hummingbird
C) Anna's hummingbird
D) Ruby-throated hummingbird

69. The United States has two national animals: The bald eagle is the national bird. What is the national mammal?
A) Bighorn sheep
B) Northwestern moose
C) Grizzly bear
D) American bison

ANSWERS ON PAGE 103

70. There are 18 species of penguins. All of the following are real penguin species; can you identify which one is the largest penguin species?
A) Little penguin
B) Royal penguin
C) Emperor penguin
D) King penguin

71. All of the following animals hibernate, which means that they go into a dormant state during the winter. Which of the following animals hibernates for the longest amount of time, up to eight months?
A) European hedgehog
B) Arctic Ground Squirrel
C) Himalayan brown bear
D) Alpine marmot

72. The axolotl is a critically endangered amphibian that is native to Mexico. Three of the following statements are true about the axolotl. Which of the following is *not* true of axolotl?
A) They are carnivorous and will eat meat.
B) They do not grow teeth.
C) They can regenerate lost body parts.
D) They like to live in trees.

73. What is the most popular pet in the world?
A) Bird
B) Snake
C) Dog
D) Cat

## 74. Which animal has the fastest heartbeat on Earth?
A) Hummingbird
B) Gecko
B) Cheetah
D) Scorpion

## 75. Some animals travel long distances, usually during specific seasons, called "migrations." Three of the following animals usually migrate. Which of the following animals does *not* usually migrate?
A) Northern cardinal
B) Flamingos
C) Salmon
D) Ruby-throated hummingbird

## 76. What is the largest species of whale in the world?
A) Beluga whale
B) Blue whale
C) Humpback whale
D) Gray whale

## 77. A mountain lion is known by multiple names, but they all refer to the same large cat. Three of the following are other names for the mountain lion. Which of the following is *not* another name for a mountain lion, but an entirely different species of cat?
A) Puma
B) Cougar
C) Panther
D) Jaguar

ANSWERS ON PAGES 103-104

78. Some animals can produce their own light. This is called "bioluminescence." Three of the following animals are bioluminescent. Which of the following animals is *not* bioluminescent?
A) Fireflies
B) Flashlight fishes
C) Lanternfish
D) Lightfish

79. Some animals are able to control and maintain the body temperature they need. These kinds of animals are known as "warm-blooded" animals, but what is the scientific term for this?
A) Invertebrates
B) Ectothermic animals
C) Endothermic animals
D) Amphibians

80. Groups of animals are sometimes given a unique name, such as a "school" of fish. Three of the group names below are true; which one is *not* a real name for a specific group of animals?
A) A "prickle" of porcupines
B) A "robbery" of wolves
C) A "troop" of kangaroos
D) A "murder" of crows

81. Sea sponges might look like plants, but they are actually animals. Three of the following statements about them are true. Which of the following statements about them is false?

A) They have microscopic mouths for eating.
B) There are more than 8,000 species of sea sponges.
C) Some species are capable of moving.
D) According to fossil evidence, they are the animal that's been on Earth the longest.

**82. All of the following animals have been or currently are considered critically endangered, which means they are in danger of becoming extinct. Three of the following animals are still considered critically endangered. Which of the following animals is no longer considered critically endangered?**
A) Giant panda
B) Black rhino
C) Bornean orangutan
D) Amur leopard

**83. Some animals have a surprising number of body parts! Three of the following statements about animals are true. Can you pick out which of the following statements is false?**
A) Horses have three stomachs.
B) Honeybees have five eyes.
C) An octopus has three hearts.
D) Lobsters have ten legs.

**84. How many stomachs does a giraffe have?**
A) 1
B) 2
C) 3
D) 4

ANSWERS ON PAGE 104

**85. While all snakes have teeth, not all snakes have fangs. All of the following snakes do have fangs. Which one has the longest fangs, measuring about 1.5-inches long, about the length of your finger?**
A) Black mamba
B) Gaboon viper
C) King cobra
D) Western diamondback rattlesnake

**86. Scientists classify animals in different ways, including whether they are "vertebrates" or "invertebrates." What does this mean?**
A) Vertebrates are animals that live in water, and invertebrates are animals that live on land.
B) Vertebrates are animals that walk vertically (or upright) on two legs, and invertebrates are animals that walk on four legs.
C) Vertebrates are animals that breathe air, and invertebrates are animals that breathe underwater.
D) Vertebrates are animals that have a backbone, and invertebrates are animals that do not.

**87. What is one major difference between amphibians and reptiles?**
A) Amphibians can live both in water and on land, while reptiles mainly live on land.
B) Amphibians breathe through gills, and reptiles are animals that breathe with lungs.
C) Amphibians have backbones, and reptiles do not.
D) Amphibians hibernate in the winter, and reptiles migrate.

**88. Which animal is known for producing cube-shaped poop?**

A) Penguin

B) Gorilla

C) Wombat

D) Kangaroo

**89. Not all animals' blood is red. Three of the following animals have blue blood. Which one has *green* blood?**

A) Octopus

B) Squid

C) Skink

D) Horseshoe crab

**90. Three of the following are different words for animal poop; which of the following is *not*?**

A) Scat

B) Dung

C) Discards

D) Droppings

**91. Which of the following is the main food that panda bears eat?**

A) Bamboo

B) Pinecones

C) Fish

D) Monkeys

ANSWERS ON PAGE 104

92. Dogs evolved from a fox-sized animal called a Leptocyon, which lived about five million years ago. Three of the following animals also evolved from the Leptocyon; which of the following did *not*?
A) Jackal
B) Hyena
C) Fox
D) Coyote

93. Bird's eggs can come in a variety of colors, including white, gray, red, brown, green and blue. Three of the following birds lay blue eggs. Which one of the following does *not* lay blue eggs?
A) Robins
B) Canaries
C) Blue jays
D) Sparrows

94. Dogs are not the only animals that bark. Three of the following animals also bark. Which one does *not*?
A) Capybara
B) Otters
C) Zebras
D) Seals

95. Some animals have multiple organs. Which of the following has nine brains?
A) Octopus
B) Earthworm
C) Alligator
D) Vulture

**96. There are only three known lizards that are poisonous. Those three are listed below. Which of the following is *not* one of the three known poisonous lizards?**
A) Gila monster
B) Mexican bearded lizard
C) Komodo dragon
D) Thorny devil

**97. Which of the following is the only snake that makes a nest when it lays its eggs?**
A) King cobra
B) Black mamba
C) Queen snake
D) Ball python

**98. Three of the following animals have fangs. Which animal in this list does *not* have fangs?**
A) Tarantula
B) Hippo
C) Baboon
D) Polar bear

**99. Spiders have eight legs, but so do other animals. Three of the animals in the following list also have eight legs. Can you pick which animal in the following list has ten legs instead of eight?**
A) Crab
B) Shrimp
C) Scorpion
D) Tick

ANSWERS ON PAGE 104

100. Birds are not the only animals that have beaks. Other animals do, too! Three of the animals in the following list have beaks; which one in this list does *not* have a beak?

A) Squid
B) Turtle
C) Toad
D) Octopus

# PLANTS TRIVIA

**101. Which of the following is the correct definition for what makes something a plant?**
A) Plants are living things that are green.
B) Plants are organisms that use energy from sunlight to make food.
C) Plants are organisms that grow out of the ground.
D) Plants are living things that can be eaten.

**102. Plants use sunlight to help them grow. There is a green substance inside plants that helps them absorb sunlight and turn it into energy. What is this substance called?**
A) Chlorophyll
B) Chlorine
C) Chronoid
D) Chronicle

ANSWERS ON PAGE 104

## 103. Which of the following is *not* considered a plant?
A) Mushroom
B) Fern
C) Cactus
D) Moss

## 104. Water evaporates from the pores in plants' leaves, similarly to how people sweat and lose water. In plants, what is this process called?
A) Expiration
B) Dilution
C) Respiration
D) Transpiration

## 105. The part of the plant that helps support and anchor the plant as well as absorb water and plant nutrients is called what?
A) Stem
B) Leaves
C) Flower
D) Roots

## 106. Plants use energy from the _____ to make food.
A) Soil
B) Rain
C) Sun
D) Wind

**107. Photosynthesis is how plants make their own**
_____.

A) Water
B) Food
C) Oxygen
D) Sunlight

**108. Plants change water and carbon dioxide into different kinds of sugars. One kind of sugar is starch, which the plant stores for later use in parts of the plant known as "starch storage organs." Three of the following vegetables that we eat are actually these "starch storage organs." Which of the following is *not* a starch storage organ?**

A) Ginger
B) Potato
C) Onion
D) Cucumber

**109. Different plants have different life cycles. Some plants are able to regrow for multiple years, while others live only for one growing cycle. What are plants called that live only for one year?**

A) Perennials
B) Biennials
C) Annuals
D) Seasonals

ANSWERS ON PAGE 104

**110. Plants need carbon dioxide, which they "breathe" in through pores on which of the following part(s) of the plant?**
A) Leaves
B) Roots
C) Stem
D) Flowers

**111. Plants give off _____ that humans need to breathe.**
A) Carbon dioxide
B) Hydrogen
C) Oxygen
D) Nitrogen

**112. Trees are often grouped into two main categories: deciduous trees and coniferous trees. What is one of the main differences between these two types of trees?**
A) Coniferous trees produce nuts.
B) Coniferous trees grow needles that stay green in winter.
C) Deciduous trees live for 100 years or more.
D) Deciduous trees need fresh water.

**113. Many plants produce a sweet liquid to attract pollinators, like bees. What is this sweet liquid called?**
A) Pollen
B) Syrup
C) Nectar
D) Honey

**114. Many plants have male and female parts in their flowers. What is the male part of the flower called?**
A)  Stamen
B)  Stalk
C)  Sickle
D)  Straddle

**115. A pollinator often helps take pollen from the male part of a flower to the female part of the flower. What is the female part of the flower called?**
A)  Pendle
B)  Pistil
C)  Perrable
D)  Pindria

**116. Three of the following statements are true about pollinators. Which statement is _not_ true?**
A)  Pollinators move pollen from one plant to another.
B)  Pollinators can be birds, bugs, slugs and even bats!
C)  Pollinators plant seeds.
D)  Honeybees are the primary pollinator of all commercial crops in the United States.

**117. Which of the following is the fastest growing land plant in the world?**
A)  Ivy
B)  Grass
C)  Bamboo
D)  Dandelions

ANSWERS ON PAGE 104

118. Some plants create a sweet smell to attract pollinators. Others produce stinky smells to attract different pollinators. Three of the following are real plants that are known for their stinky smells; which one does *not* create a stinky smell?
A) Corpse flower
B) Carrion flower
C) Skunk cabbage
D) Cheeseweed

119. While many plants produce seeds that help them grow new plants, there are plants that do *not* produce seeds. They reproduce in different ways, such as producing spores. Which of the following is one of these non-seed producing plants?
A) Moss
B) Poison ivy
C) Grass
D) Vines

120. Three of the following are ways that seeds naturally disperse themselves so they can get planted. Which of the following is *not* one of the ways seeds get themselves planted in nature?
A) By being eaten by animals
B) By being blown by the wind
C) By being carried by passing animals' fur
D) By being dissolved by rainfall

121. The oldest living group of plants in the world is estimated to be about 200,000 years old. What kind of plant is it, and where is it found?

A) Quaking aspens in a Utah forest, known as "Pando"
B) Seagrass in the Mediterranean Sea
C) Mosses found at the top of Mount Everest
D) Cacti in the Sahara Desert

122. There is one tree that scientists call "a living fossil" because it has been growing on Earth for more than 270 million years and still grows in nature today. It has remained relatively unchanged for about 80 million years and has no known relatives. Which tree is this "living fossil"?

A) Gingko
B) Sycamore
C) Juniper
D) Magnolia

123. Coconuts often wash out to sea, allowing the seeds inside to be planted all around the world! Three of the following characteristics are true about coconuts and how they can travel these long distances. Which of the following statements is *not* true about coconuts?

A) Their hairy outsides are waterproof.
B) It takes up to 220 days for a coconut to sprout.
C) The coconuts are partially hallow inside, allowing them to float.
D) They require being soaked in saltwater before being planted.

ANSWERS ON PAGE 104

**124. The roots of some most plants do not need to grow in soil, including a group of plants known as "epiphytes." Where do their roots grow instead?**
A) In the water
B) In the snow
C) In outer space
D) In the air

**125. Botanically, a fruit is considered a berry if it grows from a single ovary in a flower. Three of the following are true berries. Which one is *not*?**
A) Banana
B) Strawberry
C) Cranberry
D) Eggplant

**126. Near which part of the potato plant will you find potatoes growing?**
A) Roots
B) Stem
C) Fruit
D) Leaves

**127. Which of the following plants naturally has thorns to protect its fruits?**
A) Strawberry
B) Blueberry
C) Blackberry
D) Cranberry

128. Chocolate comes from the cacao tree. Which part of the tree do people use to make chocolate?
A) Leaves
B) Beans
C) Roots
D) Stem

129. All green plants make sugar, but there are special ones people harvest for their sugar, which can then be used in cooking and baking. Three of the following are plants humans often harvest for sugar. Which one is *not*?
A) Maple tree
B) Sugar beet
C) Date palm
D) Sugarplum tree

130. Three of the following are technically fruits. Which one is a true vegetable?
A) Cucumber
B) Onion
C) Tomato
D) Green beans

131. Native Americans often planted three of the following plants together in their gardens. Which one of the following is *not* one of the three they grew?
A) Corn
B) Beans
C) Carrots
D) Squash

ANSWERS ON PAGE 104

**132. Multiple vegetables that we commonly eat today originally came from one specific plant (*Brassica oleracea*), though now they don't resemble each other. Three of the following vegetables all came from that original plant. Which one did *not*?**
A) Kale
B) Broccoli
C) Cabbage
D) Green beans

**133. Which of the following is the only fruit that has its seeds on the outside of its skin?**
A) Strawberry
B) Raspberry
C) Banana
D) Pineapple

**134. Vanilla, a flavoring used to make things like vanilla ice cream, comes from the dried seed pods of which kind of plant?**
A) Cacao
B) Orchid
C) Rose
D) Milkweed

**135. One kind of tree is struck by lightning more than any other kind of tree. Which tree is it?**
A) Oak
B) Walnut
C) Maple
D) Pine

**136. Carrots were originally not orange. What color were they originally?**
A) Green
B) Yellow
C) Purple
D) Blue

**137. "Banana" comes from the Arabic word for which of the following?**
A) Yellow
B) Sweet
C) Finger
D) Monkey

**138. During the 1600s, one type of plant became so popular in Holland that it was worth more than gold. What plant was it?**
A) Tulips
B) Roses
C) Dandelions
D) Goldenrod

**139. It turns out that many plants whose fruit we eat are part of the rose family, technically known as the "Rosaceae" family. Three of the following fruits are part of this family; which of the following is *not*?**
A) Peach
B) Pear
C) Strawberry
D) Blueberry

ANSWERS ON PAGE 104

140. Acorns are the nuts that come from oak trees. However, oak trees do not start producing acorns until they are at least 20 years old. How old are most oak trees when they produce the most acorns?
A) 20 to 50 years old
B) 50 to 80 years old
C) 80 to 110 years old
D) 110 to 140 years old

141. The mallow family includes about 1,500 species of herbs, shrubs and trees (usually with five-petalled flowers) that humans have used for a wide variety of uses, including making marshmallows.

Three of the following are other products made from plants in the mallow family. Which of the following products does *not* come from the mallow family?
A) Okra
B) Cotton
C) Chocolate
D) Almonds

142. Durian is a tropical fruit from Asia that has a spiky exterior. It is also known for its strong smell. What is its smell frequently compared to?
A) Rotten onions and raw sewage
B) Rose and hibiscus flowers
C) Peppermint and tea
D) Basil and spices

143. Whether to protect themselves or to attract pollinators, some plants disguise themselves to look like other things. Three of the following examples are real ways plants do this; which of the following is *not*?
A)  An orchid looks like a flower with a female bee resting on top to attract male bees.
B)  A bush grows flowers that look like berries to attract certain hummingbirds.
C)  A succulent that looks like stones to avoid being eaten by hungry animals.
D)  A carnivorous plant grows what looks like raindrops but really are sticky traps for bugs.

144. Some plants can fight back when they sense they are being eaten by animals. Tomatoes do this when caterpillars begin eating their leaves. What does the tomato plant do to stop the caterpillars from eating more leaves?
A)  They produce and release poison.
B)  They grow thorns.
C)  They grow extra leaves.
D)  They trap the caterpillars in tomato juice.

145. A carnivorous plant traps and eats insects or small animals. Three of the following are carnivorous plants. Which one is *not* a carnivorous plant?
A)  Fig tree
B)  Venus flytrap
C)  Butterwort
D)  Snakeberry

ANSWERS ON PAGE 104

146. When farmers want to harvest this kind of fruit, they flood their fields with water because this fruit has air pockets inside that allow it to float on the water. Which fruit is it?
A) Strawberry
B) Blueberry
C) Cranberry
D) Cherry

147. A variety of plants are harvested for how they smell and used to make fragrances. Three of the following are harvested for their scents; which of the following is *not*?
A) Sandalwood
B) Frankincense
C) Daffodil
D) Lavender

148. Which of the following is the largest fruit that comes from a tree and can weigh up to 80 pounds?
A) Coconut
B) Jackfruit
C) Pineapple
D) Fig

149. This plant only flowers about once every 50 to 100 years. When that happens, all plants of this type will flower at the exact same time. What kind of plant is it?

A) Corpse plant
B) Bamboo
C) Orchid
D) Passionflower

**150. Banyan trees are a kind of fig tree and are considered the world's largest trees in terms of the area each one covers—a single tree in India covers more than 4 acres! Which of the following statements accurately explains how are banyan trees are able to grow so large?**
A) Their roots grow down from its branches.
B) Their leaves are very large.
C) Their trunks are made of a very strong wood.
D) Their branches twist and grow along the ground like vines.

ANSWERS ON PAGE 104

# ARE YOU HAVING FUN WITH SCIENCE?!

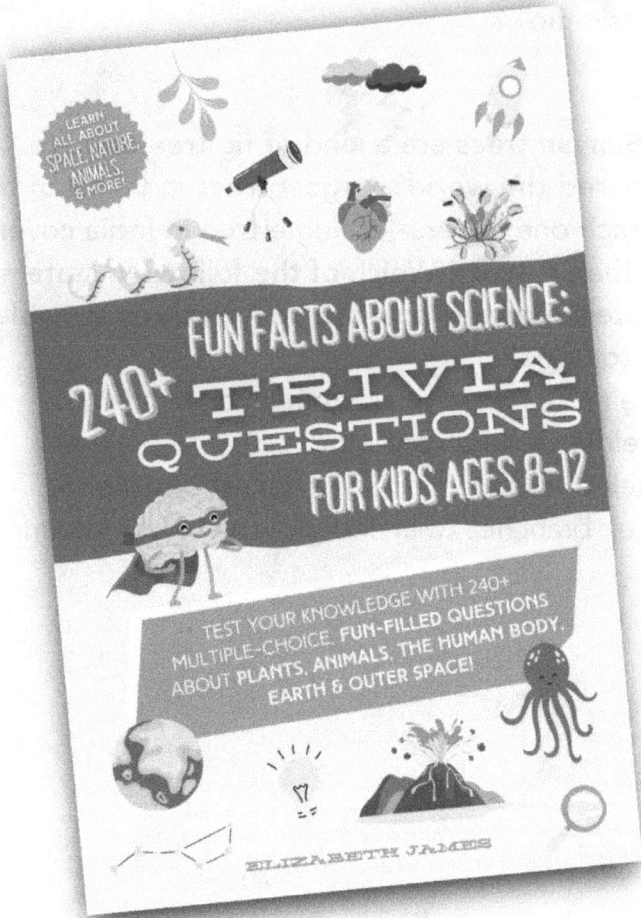

# LEAVE A REVIEW!

## THEY REALLY HELP!

# EARTH SCIENCE TRIVIA

**151. There are seven continents on Earth. Which of the following is Earth's largest continent?**
A) North America
B) Asia
C) Africa
D) Antarctica

**152. Which continent is Earth's smallest continent?**
A) Australia
B) Europe
C) South America
D) North America

**153. Which of the following is the largest island on Earth?**
A) Madagascar
B) Great Britain
C) Greenland
D) Iceland

ANSWERS ON PAGES 104

**154. Which of the following rivers is the longest river on Earth?**
A) Mississippi River
B) Nile River
C) Congo River
D) Yellow River

**155. Which of the following is the world's largest ocean?**
A) Atlantic Ocean
B) Indian Ocean
C) Pacific Ocean
D) Arctic Ocean

**156. What do we call large, mostly flat areas in nature that don't have many trees?**
A) Plains
B) Mountains
C) Plateaus
D) Valleys

**157. We normally talk about four seasons: winter, spring, summer and fall. However, in the tropics, located near the equator, they designate two seasons. What are these two seasons called?**
A) Wet and dry
B) Hot and cold
C) Sun and shade
D) Indoor and outdoor

**158. What do we call a dry region that receives little rain and where few plants grow?**
A) Oasis
B) Delta
C) Desert
D) Rainforest

**159. Three of the following are often found in or around caves. Which of the following is *not* associated with caves?**
A) Sinkhole
B) Stalagmite
C) Stalactite
D) Dike

**160. Which of the following can people use to help them predict weather?**
A) Trees
B) Moon
C) Stars
D) Clouds

**161. "Cirrus" clouds are thin, wispy tufts of clouds found high in the sky and are usually associated with fair weather. What are these clouds made of?**
A) Wind
B) Ice crystals
C) Warm water
D) Hydrogen

ANSWERS ON PAGE 104-105

**162. White fluffy clouds with lots of blue sky between them are called "cumulus" clouds. What kind of weather is associated with these kinds of clouds?**
A) Sunny
B) Rainy
C) Hot
D) Foggy

**163. Thick clouds that hang low and cover the sky are known as "stratus" clouds. Which of the following weather conditions might be associated with these kinds of clouds?**
A) Hail
B) Snow
C) Fog
D) Wind

**164. When meteorologists use the word "precipitation," what are they talking about?**
A) Mountains, hills, cliffs, etc.
B) Different kinds of earthquakes
C) Lightning and thunder
D) Rain, hail, sleet, snow, etc.

**165. Glaciers are large sheets of ice. Where are most of the world's glaciers found?**
A) Alaska
B) North Pole
C) Antarctica
D) Iceland

**166. Tornadoes are funnels of rapidly rotating air during a thunderstorm. Three of the following can be signs to let you know that a tornado might develop. Which of the following is *not* a typical sign that a tornado is likely to develop?**
A)  Thunderclouds with bulging lumps, known as "mammatus" clouds
B)  Cool and chilly weather
C)  The sky is an eerie color, such as brown, yellow or green
D)  Large hail without any rain

**167. Which country experiences the most tornadoes in the world?**
A)  United States
B)  Brazil
C)  Russia
D)  Germany

**168. Hurricanes are tropical cyclones that occur in areas near oceans. There are seven "basins" where these hurricanes are concentrated. Which of the following basins receives more hurricanes than the other basins?**
A)  Atlantic basin
B)  Southwest Indian basin
C)  Australian/Southwest Pacific basin
D)  Northwest Pacific basin

ANSWERS ON PAGE 105

169. Thunderstorms are heavy storms with lightning and thunder. While they can occur any time of year, during which two seasons can you expect to experience the most thunderstorms?
A) Fall and winter
B) Winter and spring
C) Spring and summer
D) Summer and fall

170. According to Guinness World Records, the largest natural _____ was 15 inches wide and 8 inches thick, found in Montana. What was it?
A) Hail
B) Snowflake
C) Icicle
D) Raindrop

171. Atmospheric pressure is pressure caused by the weight of the atmosphere. Sometimes a change in pressure can affect the weather. For instance, it might rain if the pressure does which of the following?
A) Rises
B) Falls
C) Jumps up and down
D) Stays the same

172. Coral reefs are found in the ocean and can come in a variety of colors. How are they made?
A) Lava from volcanoes that has hardened
B) Dead algae

C) Rocks that have grown on the ocean floor
D) Small living organisms called polyps

**173. Did you know that visible light is not white? It is actually made up of 7 different colors. Three of the following colors are found in visible light; which of the following is *not* one of the seven colors found in visible light?**
A) Green
B) Violet
C) Brown
D) Blue

**174. When light hits an object, some of the light can be absorbed by or reflected off the object. If an object *absorbs* all the colors in the visible spectrum, what color will that object appear to be?**
A) White
B) Black
C) Blue
D) Yellow

**175. When light hits an object, some of the light can be absorbed by or reflected off the object. If an object *reflects* all of the colors in the visible spectrum, what color will that object appear to be?**
A) White
B) Black
C) Blue
D) Yellow

ANSWERS ON PAGE 105

176. Minerals are formed when certain chemicals combine together and harden, forming crystals. Three of the following are examples of minerals. Which of the following is not considered a mineral but a rock?
A) Quartz
B) Ruby
C) Copper
D) Granite

177. Rocks are different from minerals because they are the combination of various minerals, all packed tightly together. There are three main types of rocks. Can you pick the one in the list below that is *not* one of the three main types of rocks?
A) Igneous rocks
B) Sedimentary rocks
C) Metamorphic rocks
D) Fossils

178. Over time, rocks break down as the result of weathering and erosion, creating different kinds of sediments. Three of the following are the most common kinds of sediment. Which one of the following is *not* a type of sediment at all?
A) Gravel
B) Soil
C) Sand
D) Clay

**179. The Earth is constantly changing and new landforms are slowly being formed by different forces. Three of the following are forces that can change landforms. Which of the following does *not* change Earth's landforms?**

A) Water and ice
B) Magma
C) Shifting tectonic plates
D) Clouds

**180. Earth's continents do not stay in one place, but slowly move over time. What do scientists call this process?**

A) Continental drift
B) Earth shifting
C) The Great Rearrangement
D) Earth momentum

**181. The earth is composed of four distinct layers: The outside layer on which we live is known as the crust. Three of the following are the other three layers. Which one in the list below is *not* one of the earth's other three layers?**

A) Inner core
B) Outer core
C) Magma layer
D) Mantle

ANSWERS ON PAGE 105

**182. The earth's crust is made up of multiple pieces that fit together like a moving jigsaw puzzle. What are these pieces called?**

A) Tectonic plates
B) Continents
C) Glaciers
D) Countries

**183. If you dig down below the soil, which of the following is the correct term for the layer you would reach next?**

A) Bedrock
B) Mineral layer
C) Fossil layer
D) Exosphere

**184. The majority of the Earth's volcanoes—452!—are found in one large area, where a number of tectonic plates meet. What is this area commonly called?**

A) Lava Loop
B) Volcano Valley
C) Ring of Fire
D) Explosion Alley

**185. Faults are places where tectonic plates have cracked and broken. Earthquakes often occur around these faults. Three of the following are the most common kinds of faults found in nature. Which one is *not* a real kind of fault?**

A) Normal fault
B) Reverse fault

C) Strike-slip fault
D) Abnormal fault

**186. Ocean levels rise and fall every day, resulting in what we call "tides." Which of the following causes these tidal changes?**
A) Whales migrating
B) Gravitational pull from the moon and sun
C) Volcanoes erupting
D) Tropical storms

**187. There is a new full moon once each month. Different traditions have given each of these full moons a special name, depending on when they occur. Three of the following are real names given to different full moons; which of the following is *not* one of the names for a specific full moon?**
A) Worm moon
B) Pink moon
C) Strawberry moon
D) Fox moon

**188. Gravity is an invisible force that pulls on objects. Three of the following statements about gravity are true. Which one is *not* true about gravity?**
A) Gravity cannot push.
B) Black holes have the strongest gravity of anything in the universe.
C) The sun does not have any gravity.
D) The force of attraction between you and the Earth is your weight.

ANSWERS ON PAGE 105

**189. What is the imaginary line called that runs through the center of a planet on which the planet turns?**
A) Equator
B) Axis
C) Latitude
D) Longitude

**190. The sun rises in the east and sets in the west. Therefore, in which direction does the Earth rotate?**
A) West to east
B) East to west
C) North to south
D) South to north

**191. The "Northern Lights" (also called "Aurora Borealis") are a natural phenomenon that look like colorful lights dancing through the sky in places like Alaska and Canada. What causes them?**
A) A special kind of cloud formation
B) City lights reflected in the air
C) Solar wind particles colliding with Earth's atmosphere
D) Raindrops freezing in the sky

# SPACE TRIVIA

**192. Which of the following objects in our solar system is the largest?**
A) Sun
B) Moon
C) Earth
D) Jupiter

**193. How many planets are in our solar system?**
A) 7
B) 8
C) 9
D) 10

**194. Which planet is the largest in our solar system?**
A) Saturn
B) Uranus
C) Jupiter
D) Earth

ANSWERS ON PAGE 105

## 195. Which planet is closest to Earth?

A) Venus
B) Mars
C) Mercury
D) Jupiter

## 196. What is the smallest planet in our solar system?

A) Neptune
B) Uranus
C) Mercury
D) Mars

## 197. Which planet in our solar system is called the "Blue Planet"?

A) Earth
B) Uranus
C) Neptune
D) Mars

## 198. Which planet is called "the King of Planets"?

A) Earth
B) Mars
C) Venus
D) Jupiter

## 199. Which planet is known as the "Red Planet"?

A) Mercury
B) Mars
C) Venus
D) Uranus

200. Which planet is known for its iconic "Great Red Spot," which is actually a giant storm?
A) Mars
B) Saturn
C) Mercury
D) Jupiter

201. Which planet in our solar system looks blue because of methane gas?
A) Earth
B) Mars
C) Neptune
D) Mercury

202. Which of the following planets was the first planet to be discovered using a telescope?
A) Mercury
B) Uranus
C) Neptune
D) Jupiter

203. The planets closest to the sun (Mercury, Venus, Earth, Mars) form what's called "the inner planets" of our solar system. They are all mostly made up of which of the following?
A) Ice
B) Gas
C) Liquid
D) Rock

ANSWERS ON PAGE 105

**204.** In 2023, researchers found that Saturn has 145 moons—making it the planet with the most moons in our solar system. Three of the following are actual names for some of its moons. Can you guess which of the following is *not* a name for one of its moons?
A) Prometheus
B) Penelope
C) Phoebe
D) Pandora

**205.** Saturn is known for its rings. However, other planets in our solar system—including three below— also have rings. Which one does *not* have rings?
A) Neptune
B) Jupiter
C) Uranus
D) Venus

**206. What are Saturn's rings made from?**
A) Lava and fire
B) Ice and rocks
C) Clouds
D) Comets

**207.** Earth is not the only planet that experiences natural disasters. All of the following have been witnessed in our solar system except _____?
A) Dust storms on Mars
B) Volcanoes on Jupiter
C) Tornadoes on Venus
D) Geysers on one of Saturn's moons

**208. The following are all dwarf planets found in our solar system. Which one used to be classified as a planet?**
A) Pluto
B) Eris
C) Ceres
D) Haumea


**209. Volcanoes can be found on multiple different planets. On which planet will you find the largest volcano in our solar system?**
A) Earth
B) Venus
C) Mars
D) Mercury


**210. Earth is not the only place in the universe with snow. Scientists have discovered snow on three of the following places in our solar system; on which have they *not* found snow?**
A) Saturn
B) Mars
C) Mercury
D) Pluto


**211. What is the name of our galaxy?**
A) Blue Galaxy
B) Milky Way
C) Solar System
D) Cereal Galaxy

ANSWERS ON PAGE 105

71

212. On any given planet, one day is the length of time it takes that planet to spin on its axis one time. On earth, that process takes 24 hours. Which planet in our solar system has the shortest day—lasting only 10 hours long?
A) Neptune
B) Jupiter
C) Uranus
D) Venus

213. Which planet in our solar system has the longest day, taking it more than four months to make one full rotation on its axis?
A) Saturn
B) Uranus
C) Venus
D) Mars

214. One year on Earth equals 365 days, because that's how long it takes Earth to go around the sun. Which planet in our solar system has the shortest year, orbiting the sun in only 88 days (in Earth time)?
A) Saturn
B) Uranus
C) Mercury
D) Mars

215. Which planet has the longest year in our solar system, taking it more than 164 years (in Earth time) to orbit the sun once?!

A) Neptune
B) Jupiter
C) Uranus
D) Venus

**216. Which star is closest to Earth?**
A) Vega
B) Sirius
C) Acrux
D) Sun

**217. Stars are giant balls of gas that generate light and heat. Their light can appear in different colors—including three of the following—depending on how much heat they emit. Which of the following is *not* a color that stars can appear as?**
A) Green
B) Blue
C) Yellow
D) Red

**218. Our universe is made up of different things, many of which we can see like stars and planets. However, there are other things in the universe that we *cannot* see. What is the term scientists have for the things in the universe that we *cannot* see?**
A) Dark matter
B) Black holes
C) Nebulae
D) Asteroids

ANSWERS ON PAGE 105

**219. Three of the following statements are true about stars; which one is *false* about stars?**
A) They can be different colors.
B) They can explode.
C) The biggest stars last the longest.
D) They are all made from the same material.

**220. How many stars can you see without using a telescope when you look up at the night sky?**
A) 200 to 500
B) 2,000 to 2,500
C) 5,000 to 10,000
D) 20,000 to 25,000

**221. A constellation is a collection of _____ that form images in the night sky.**
A) Comets
B) Stars
C) Galaxies
D) Satellites

**222. There are 88 recognized constellation names. Which of the following is not one of them?**
A) Feline (cat)
B) Aquila (eagle)
C) Cancer (crab)
D) Lacerta (lizard)

**223. Many of the constellations get their names from:**
A)  Fairytales
B)  The Bible
C)  Old movies
D)  Greek mythology

**224. What's it called when a very large star explodes?**
A)  A black hole
B)  A nebula
C)  A supernova
D)  A solar flare

**225. What is the scientific name for our sun?**
A)  Suni
B)  Sol
C)  Star #1
D)  Helios

**226. A "shooting star" is actually a _____.**
A)  Comet
B)  Satellite
C)  Wish
D)  Meteor

**227. Three of the following are the three main types of galaxies. Which one is *not* a kind of galaxy?**
A)  Spiral
B)  Elliptical
C)  Irregular
D)  Curved

ANSWERS ON PAGE 105

**228. Researchers have discovered more than 50 galaxies in the universe. Three of the following are names of real galaxies; which of the following is *not*?**
A) Peekaboo Galaxy
B) Cartwheel Galaxy
C) Backward Galaxy
D) Starburst Galaxy

**229. A satellite is something in space that orbits something else. Satellites can be natural or man-made, such as those put into space to help provide Internet. Which of the following is a natural satellite orbiting around the Earth?**
A) Moon
B) Mars
C) Sun
D) Asteroid belt

**230. Three of the following facts are true about the moon. Which of the following statements is *false* about the moon?**
A) It makes no light of its own.
B) The moon is smaller than the Earth.
C) There is no gravity on the moon.
D) There is no wind or water on the moon.

**231. When the moon is between the sun and the Earth, it is completely in the shadows. What do we call this lunar phase that the moon experiences about every 28 days?**

A) New moon
B) Waning gibbous
C) Waxing crescent
D) Full moon

**232. The Earth moves around the Sun while the moon moves around Earth. What is the phenomenon called during a new moon when the moon casts its shadow on the Earth?**
A) Lunar eclipse
B) Solar eclipse
C) Blackout
D) Full moon

**233. In space, what is it called when gravity is so strong that nothing—not even light—can escape its gravitational pull?**
A) A nebula
B) A supernova
C) A black hole
D) A solar flare

**234. Three of the following statements are true about black holes. Which statement is *false*?**
A) They are invisible and can only be seen with special telescopes.
B) Light cannot escape them.
C) Scientists believe there are millions of them.
D) They aren't real.

ANSWERS ON PAGE 105

235. Planets are not the only things that orbit the sun. Three of the following objects also orbit the sun. Which of the following does *not* orbit the sun?
A) Nebulae
B) Comets
C) Asteroids
D) Dwarf planets

236. What famous comet orbits the sun and is visible on Earth when it passes by every 76 years?
A) Twain's Comet
B) Armstrong's Comet
C) Halley's Comet
D) Washington's Comet

237. Which of the following is the correct definition for the term "meteoroids"?
A) Rocky objects in space that revolve around the sun but are too small and numerous to be considered planets.
B) Loose collections of ice, dust or rock particles whose orbits are usually long, narrow ellipses.
C) A chunk of rock or dust in space that comes from a comet or asteroid.
D) A chunk of rock or dust that enters Earth's atmosphere and lands on Earth.

**238. In 1969, who was the first person to walk on the moon?**
A) Neil Armstrong
B) Buzz Aldrin
C) John Glenn
D) Buzz Lightyear

**239. What was the space mission called that carried the first humans to the moon in 1969?**
A) Apollo 10
B) Pioneer 10
C) Viking 1
D) Apollo 11

**240. Humans have only landed on the moon, however we have explored other parts of our solar system using unmanned robots called rovers. We have used rovers to explore all of the following except _____?**
A) Mars
B) Mercury
C) Venus
D) Saturn's moon, Titan

ANSWERS ON PAGE 105

**241. NASA is the U.S. government agency responsible for science and technology related to space and air. What does NASA stand for?**
A) National Aeronautics and Space Administration
B) North American Shuttle Authority
C) New Age Science Agency
D) Next Astrophysics Star Association

**242. Which of the following were the first living creatures to enter space in February 1947?**
A) Cockroaches
B) Tarantulas
C) Ladybugs
D) Fruit flies

**243. All of the following animals have travelled to space. Which one of the following was the first animal to fly into deep space and travel around the moon?**
A) Monkey
B) Tortoise
C) Guinea pig
D) Dog

# ARE YOU HAVING FUN WITH SCIENCE?!

LEARN ALL ABOUT SPACE, NATURE, ANIMALS, & MORE!

FUN FACTS ABOUT SCIENCE: 240+ TRIVIA QUESTIONS FOR KIDS AGES 8-12

TEST YOUR KNOWLEDGE WITH 240+ MULTIPLE-CHOICE, FUN-FILLED QUESTIONS ABOUT PLANTS, ANIMALS, THE HUMAN BODY, EARTH & OUTER SPACE!

ELIZABETH JAMES

# LEAVE A REVIEW!

THEY REALLY HELP!

# CHECK OUT MORE TRIVIA FUN!

## 240+ TRIVIA QUESTIONS ABOUT ALL YOUR FAVORITE HOLIDAYS!

# GAMES

Take the fun to the next level with one of the following eight challenge games that will put your knowledge about science to the ultimate test!

We have included options for playing by yourself as well as with multiple players or dividing into teams. Each game has its own unique spin, so we encourage you to test them all and see which one is the winner as your family's favorite!

Some involve minimal equipment—such as paper, pencil or a timer—but they are all are intended to be ultra-portable so you can even play while you're on the go, such as while waiting for your food at a restaurant or on a long car ride.

The fun doesn't ever have to stop!

# GAME #1

# THE 20 POINT CHALLENGE

*Test your knowledge by answering trivia questions correctly and being the first to reach 20 points!*

**Number of Players:** 1 or more

**Instructions:**
- Pass out a piece of paper and a writing utensil to each player.
- Determine which category you will read questions from.
- Read 10 questions out loud at a time, with all players recording their own answers at this time. (One person may read all of the questions, or players may take turns reading them out loud. If desired, use a timer for recording responses.)
- Once all 10 questions have been read, look up the answers in the back of the book. Each participant may score their own sheets, or they may exchange them for scoring purposes.
- Each correct answer equals one point.
- Play additional rounds and score those questions until one player reaches 20 points—making them the winner!

# GAME #2

# SHOWDOWN CHALLENGE

*How much does your opponent know? Find out with this game that involves selecting questions to put other teams to the test!*

**Number of Players:** 2 or more (individually or as teams)

**Instructions:**
- Determine which category you will read questions from and for how many rounds you will play.
- The first player (or team) to go can be determined by flipping a coin, playing "Rock, Paper, Scissors" or another method.
- Players will take turns selecting a question for their opponent(s), and then read that question and multiple-choice options aloud. The guessing player(s) will select an answer within a set amount of time.
- Correct responses are worth one point. Tally players' scores.
- The next player (or team) will select a question for their opponent(s), and play will continue like this until all players have taken an equal number of turns reading questions.
- The team or player with the highest score at the end of the last round wins.

# GAME #3

# RACE THE CLOCK SOLO CHALLENGE

*How many questions can you answer in a set amount of time? Take this challenge to find out!*

**Number of Players:** 1 (solo game)

**Instructions:**
- The goal in this challenge is to see how many answers you can correctly guess in the given amount of time, so select a category for which you will be answering questions and the amount of time you will give yourself.
- Once you're ready, start the timer and begin recording your answers to the questions in your selected category.
- Continue answering until the timer goes off. Score your answers and make note of how many you got correct.
- Repeat for multiple rounds with new questions, trying to beat your previous score. You can also compare your scores with someone else who's completed this challenge!

# BACKWARD TRIVIA CHALLENGE

*What if you were given the trivia **answer**—could you figure out the trivia **question**? Find out in this upside-down challenge with a 20-questions twist!*

**Number of Players:** 2 or more (individually or as teams)

**Instructions:**
- Determine which category you will read questions from and which player (or team) will start.
- The first player (or team) will select one trivia question. Then, they will check the back of the book for the correct answer and read *only* the entire correct answer to the other player(s).
- The opponent(s) must try to guess the original question for that answer. To do so, they can ask up to 20 "yes" or "no" questions, to which the player(s) that selected the question must answer.
- Keep track of how many questions the guessing player(s) have asked. They can ask up to 20 questions before they must make their final guess. If they guess correctly, they get one point. If they do not, the asking player(s) get one point.
- Switch roles and continue play until all players (or teams) have gone an equal number of times. The winner is the team with the most points at the end.

# GAME #5

# TIMER HOT SEAT CHALLENGE

*How many questions can players correctly answer in a set amount of time? This challenge will put them to the test!*

**Number of Players:** 2 or more

**Instructions:**
- Determine which category you will read questions from, and which player will go first.
- That player will select 10 questions in sequential order to ask, and write down the answers to those questions before the round begins.
- Another player will be chosen to be in the "Hot Seat" and must answer the selected questions as quickly as possible.
- When the round begins, start a timer.
- If the player answers correctly, the asking player will move on to the next question. However, if the answer is incorrect, the asking player will inform they are wrong and the player in the "Hot Seat" must guess again until they get the correct answer.
- Once the player has answered all 10 questions correctly, stop the timer and record the time.
- Change roles, and repeat until all players have had a chance in the "Hot Seat." The winner is the player or team with the lowest time.

# GAME #6

# ROLL TO WIN CHALLENGE

*Is luck on your side? Find out when you determine the value of correct answers with the roll of dice!*

**Number of Players:** 1 or more

**Instructions:**
- You will need a dice for this game, as well as a piece of paper and a writing utensil to keep score. (You can also search up "virtual dice roller" on a smartphone.)
- Determine which player will go first. That player will roll the dice, which will assign the point value he or she will receive if they gets the answer to their question correct.
- Read the question and select an answer.
- Check the answer; if correct, the player will receive as many points as he or she rolled with the dice. If not, the player receives zero points. Record the score.
- The next player will roll the dice to determine the value of a correct answer for their question. Play continues in this pattern.
- The player who receives 30 points first wins!

# GAME #7

# 3 STRIKES CHALLENGE

*How many questions can you correctly answer—without getting more than 3 incorrect? Take this challenge and find out!*

**Number of Players:** 1 or more

**Instructions:**
- The goal of this game is to see how many questions you can answer correctly—while only being able to get 3 wrong before a player is "out."
- Give each player a piece of paper and a writing utensil, and determine which category you will read questions from.
- One player will read a question, and each player will write down their own answers. (Use a timer if desired.) Afterward, read the correct answer.
- If anyone gets the answer wrong, they get one strike. Keep track of each player's strikes.
- Move on to the next question. (A new player can read, if desired.) Continue scoring after each question. When any player gets their third strike, they are "out." Play continues until there is only one player remaining, who is crowned the winner.

# GAME #8

# TRIVIA TITAN CHALLENGE

*Who really knows their stuff? The only way to find out is to put them to the test—without any multiple-choice answers! Instead, players must rely on their knowledge to answer correctly.*

**Number of Players:** 2 or more (individually or as teams)

**Instructions:**
- Determine which category you will use.
- One player will read the question and another player will answer. Determine which players will start in each of those roles.
- The player asking the question will select a question to read but will *not* read the answers.
- The answering player must come up with the answer based on their already-existing knowledge. If desired, use a timer.
- If they get the answer correct, they earn one point. If not, no points are awarded. Keep score.
- Switch roles, and ask a new question.
- Continue until all players have gone an equal amount of times. The winner with the most points at the end of the game is crowned "Trivia Titan!"

# CHECK OUT MORE TRIVIA FUN!

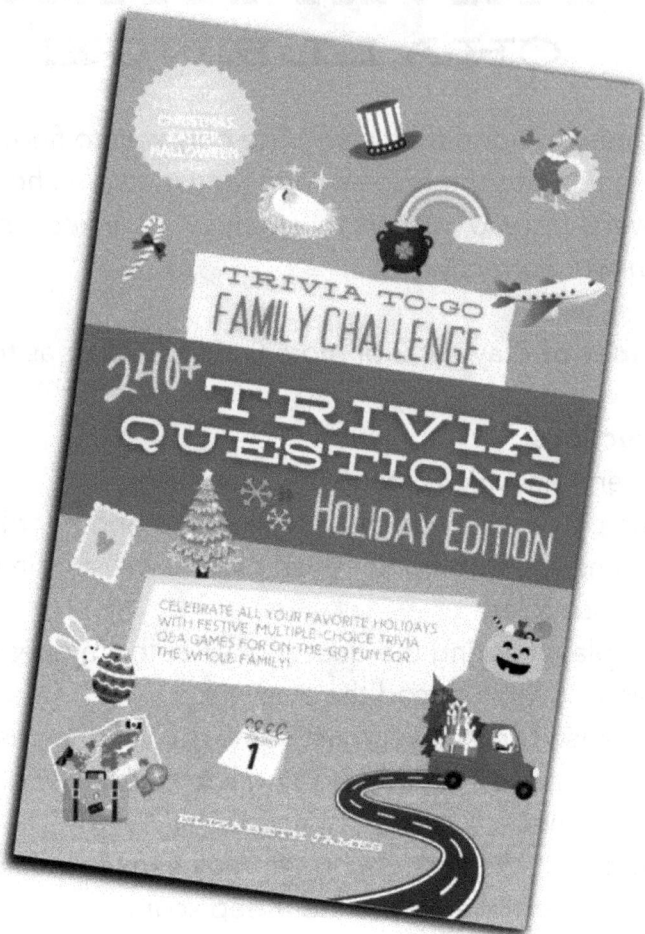

## 240+ TRIVIA QUESTIONS ABOUT ALL YOUR FAVORITE HOLIDAYS!

# ANSWERS

Please note that the answers on the following pages are intentionally arranged in a non-sequential pattern. For instance, the answers to questions 1, 11, 21, 31 and so forth are grouped together, followed by a group for questions 2, 12, 22, 32 and so on.

This deliberate arrangement is designed to prevent inadvertently seeing the answer to the next question when checking a response. We hope this ensures an exciting and engaging trivia experience while playing.

Enjoy the challenge!

# EXCLUSIVE FREEBIE!

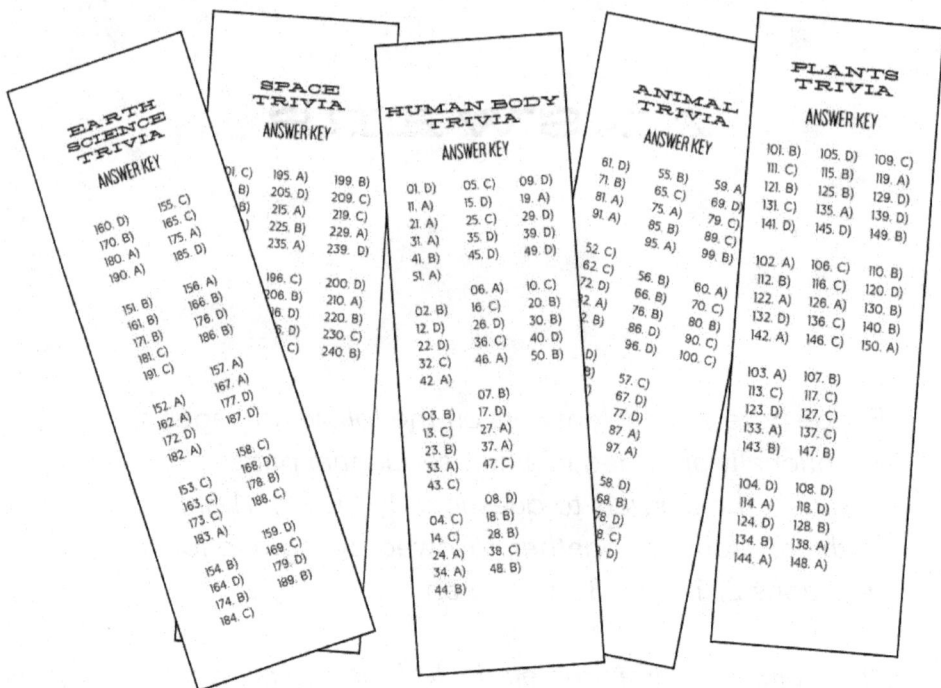

# GET ALL THE ANSWERS AT YOUR FINGERTIPS!

## DOWNLOAD FREE PRINTABLE ANSWER KEY BOOKMARKS

### WHEN YOU SIGN UP FOR EMAILS!

Visit elizabethjameswrites.com/scienceanswers

# ANSWERS
## QUESTIONS #1-79

| | | | |
|---|---|---|---|
| 01. D) | 43. C) | 06. A) | 48. B) |
| 11. A) | 53. D) | 16. C) | 58. D) |
| 21. A) | 63. B) | 26. D) | 68. B) |
| 31. A) | 73. C) | 36. C) | 78. D) |
| 41. B) | | 46. A) | |
| 51. A) | 04. C) | 56. B) | 09. D) |
| 61. D) | 14. C) | 66. B) | 19. A) |
| 71. B) | 24. A) | 76. B) | 29. D) |
| | 34. A) | | 39. D) |
| 02. B) | 44. B) | 07. B) | 49. D) |
| 12. D) | 54. A) | 17. D) | 59. A) |
| 22. D) | 64. A) | 27. A) | 69. D) |
| 32. C) | 74. A) | 37. A) | 79. C) |
| 42. A) | | 47. C) | |
| 52. C) | 05. C) | 57. C) | 10. C) |
| 62. C) | 15. D) | 67. D) | 20. B) |
| 72. D) | 25. C) | 77. D) | 30. B) |
| | 35. D) | | 40. D) |
| 03. B) | 45. D) | 08. D) | 50. B) |
| 13. C) | 55. B) | 18. B) | 60. A) |
| 23. B) | 65. C) | 28. B) | 70. C) |
| 33. A) | 75. A) | 38. C) | |

# ANSWERS
## QUESTIONS #80-159

| | | | |
|---|---|---|---|
| 80. B) | 132. D) | 85. B) | 137. C) |
| 90. C) | 142. A) | 95. A) | 147. B) |
| 100. C) | 152. A) | 105. D) | 157. A) |
| 110. B) | | 115. B) | |
| 120. D) | 83. A) | 125. B) | 88. C) |
| 130. B) | 93. D) | 135. A) | 98. D) |
| 140. B) | 103. A) | 145. D) | 108. D) |
| 150. A) | 113. C) | 155. C) | 118. D) |
| | 123. D) | | 128. B) |
| 81. A) | 133. A) | 86. D) | 138. A) |
| 91. A) | 143. B) | 96. D) | 148. A) |
| 101. B) | 153. C) | 106. C) | 158. C) |
| 111. C) | | 116. C) | |
| 121. B) | 84. D) | 126. A) | 89. C) |
| 131. C) | 94. B) | 136. C) | 99. B) |
| 141. D) | 104. D) | 146. C) | 109. C) |
| 151. B) | 114. A) | 156. A) | 119. A) |
| | 124. D) | | 129. D) |
| 82. A) | 134. B) | 87. A) | 139. D) |
| 92. B) | 144. A) | 97. A) | 149. B) |
| 102. A) | 154. B) | 107. B) | 159. D) |
| 112. B) | | 117. C) | |
| 122. A) | | 127. C) | |

# ANSWERS
## QUESTIONS #160-243

| | | | |
|---|---|---|---|
| 160. D) | 202. B) | 165. C) | 227. D) |
| 170. B) | 212. B) | 175. A) | 237. C) |
| 180. A) | 222. A) | 185. D) | |
| 190. A) | 232. B) | 195. A) | 168. D) |
| 200. D) | 242. D) | 205. D) | 178. B) |
| 210. A) | | 215. A) | 188. C) |
| 220. B) | 163. C) | 225. B) | 198. D) |
| 230. C) | 173. C) | 235. A) | 208. A) |
| 240. B) | 183. A) | | 218. A) |
| | 193. B) | 166. B) | 228. D) |
| 161. B) | 203. D) | 176. D) | 238. A) |
| 171. B) | 213. C) | 186. B) | |
| 181. C) | 223. D) | 196. C) | 169. C) |
| 191. C) | 233. C) | 206. B) | 179. D) |
| 201. C) | 243. B) | 216. D) | 189. B) |
| 211. B) | | 226. D) | 199. B) |
| 221. B) | 164. D) | 236. C) | 209. C) |
| 231. A) | 174. B) | | 219. C) |
| 241. A) | 184. C) | 167. A) | 229. A) |
| | 194. C) | 177. D) | 239. D) |
| 162. A) | 204. B) | 187. D) | |
| 172. D) | 214. C) | 197. A) | |
| 182. A) | 224. C) | 207. C) | |
| 192. A) | 234. D) | 217. A) | |

www.ingramcontent.com/pod-product-compliance
Lightning Source LLC
Chambersburg PA
CBHW070128030426
42335CB00016B/2306